SPORTS

SOCCER

by Mari Schuh

AMICUS | AMICUS INK

goal

cleats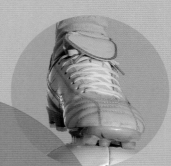

Look for these
words and pictures
as you read.

 sideline

 goalie

A player kicks the ball far.
A soccer match starts.
Let's watch!

Two teams play.
Each team has 11
players on the field.

Do you see the goal?
A player kicks the ball.
It goes into the net.
Score! One point!

goal

Do you see the cleats?
They have spikes.
They help players run fast.

cleats

Do you see the sideline?
It marks the edge of the field.
The ball went out.
A player kicks it back in.

sideline

Do you see the goalie?
She grabs the ball.
No point for the other team!

goalie

The players pass the ball.
They run fast. Go team!

Do you see the goal?
A player kicks the ball.
It goes into the net.
Score! One point!

goal

goal

Do you see the cleats?
They have spikes.
They help players run fast.

cleats

cleats

Did you find?

sideline

goalie

Do you see the sideline?
It marks the edge of the field.
The ball went out.
A player kicks it back in.

sideline

Do you see the goalie?
She grabs the ball.
No point for the other team!

goalie

Spot is published by Amicus and Amicus Ink
P.O. Box 1329, Mankato, MN 56002
www.amicuspublishing.us

Library of Congress Cataloging-in-Publication Data
Names: Schuh, Mari C., 1975- author.
Title: Soccer / by Mari Schuh.
Description: Mankato, Minnesota : Amicus, 2018. | Series: Spot.
 Sports | Audience: K to Grade 3.
Identifiers: LCCN 2016057199 (print) | LCCN 2016058337
 (ebook) | ISBN 9781681510897 (library binding) | ISBN
 9781681522081 (pbk.) | ISBN 9781681511795 (ebook)
Subjects: LCSH: Soccer--Juvenile literature. | Picture puzzles--
 Juvenile literature.
Classification: LCC GV943.25 .S36 2018 (print) | LCC
 GV943.25 (ebook) | DDC 796.334--dc23
LC record available at https://lccn.loc.gov/2016057199

Printed in China

HC 10 9 8 7 6 5 4 3 2 1
PB 10 9 8 7 6 5 4 3 2 1

To St. Paul Lutheran School — MS

Rebecca Glaser, editor
Deb Miner, series designer
Aubrey Harper, book designer
Holly Young, photo researcher

Photos by: Alamy Stock Photo/
Cal Sport Media, 3, dpa picture
alliance archive, 10–11, PA Images,
14–15; Getty Images/Blend
Images/Erik Isakson, cover, Andre
Ringuette, 6–7, Feng Li, 12–13;
iStock/Eugene_Onischenko, 1,
sbhaumik, 8–9; Shutterstock/Pavel
L Photo and Video, 4–5

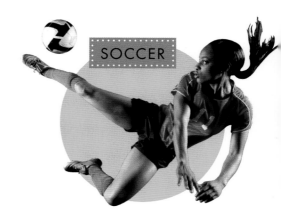

SOCCER